ACCIDENTAL
SCIENCE DISCOVERIES

PLASTIC

Kenny Abdo

Fly!
An Imprint of Abdo Zoom
abdobooks.com

abdobooks.com

Published by Abdo Zoom, a division of ABDO, P.O. Box 398166, Minneapolis, Minnesota 55439. Copyright © 2024 by Abdo Consulting Group, Inc. International copyrights reserved in all countries. No part of this book may be reproduced in any form without written permission from the publisher. Fly!™ is a trademark and logo of Abdo Zoom.

Printed in the United States of America, North Mankato, Minnesota.
102023
012024

Photo Credits: Alamy, Bridgeman Images, Getty Images, Shutterstock
Production Contributors: Kenny Abdo, Jennie Forsberg, Grace Hansen
Design Contributors: Candice Keimig, Neil Klinepier, Colleen McLaren

Library of Congress Control Number: 2023938005

Publisher's Cataloging-in-Publication Data

Names: Abdo, Kenny, author.
Title: Plastic / by Kenny Abdo
Description: Minneapolis, Minnesota : Abdo Zoom, 2024 | Series: Accidental science discoveries | Includes online resources and index.
Identifiers: ISBN 9781098284138 (lib. bdg.) | ISBN 9781098284855 (eBook) | ISBN 9781098285210 (Read-to-Me eBook)
Subjects: LCSH: Plastics--Juvenile literature. | Plastic materials--Juvenile literature. | Serendipity in science--Juvenile literature. | Inventions--Juvenile literature. | Discoveries in science--Juvenile literature.
Classification: DDC 500--dc23

TABLE OF CONTENTS

Plastic 4

The Accident 6

The Discovery 10

The Footprint 20

Glossary 22

Online Resources 23

Index 24

PLASTIC

Winning the cash prize in a contest was all one inventor wanted. Changing industries and life as people knew it was just a happy accident!

THE ACCIDENT

John Wesley Hyatt entered a contest in 1869. A billiards company wanted a substitute for **ivory** pool balls. While working, Hyatt accidentally spilled two special substances he was working with that mixed together.

Hyatt noticed that the material dried into a tough but soft film. He called the material **celluloid**. He did not win the contest. But what he did do was accidentally create plastic!

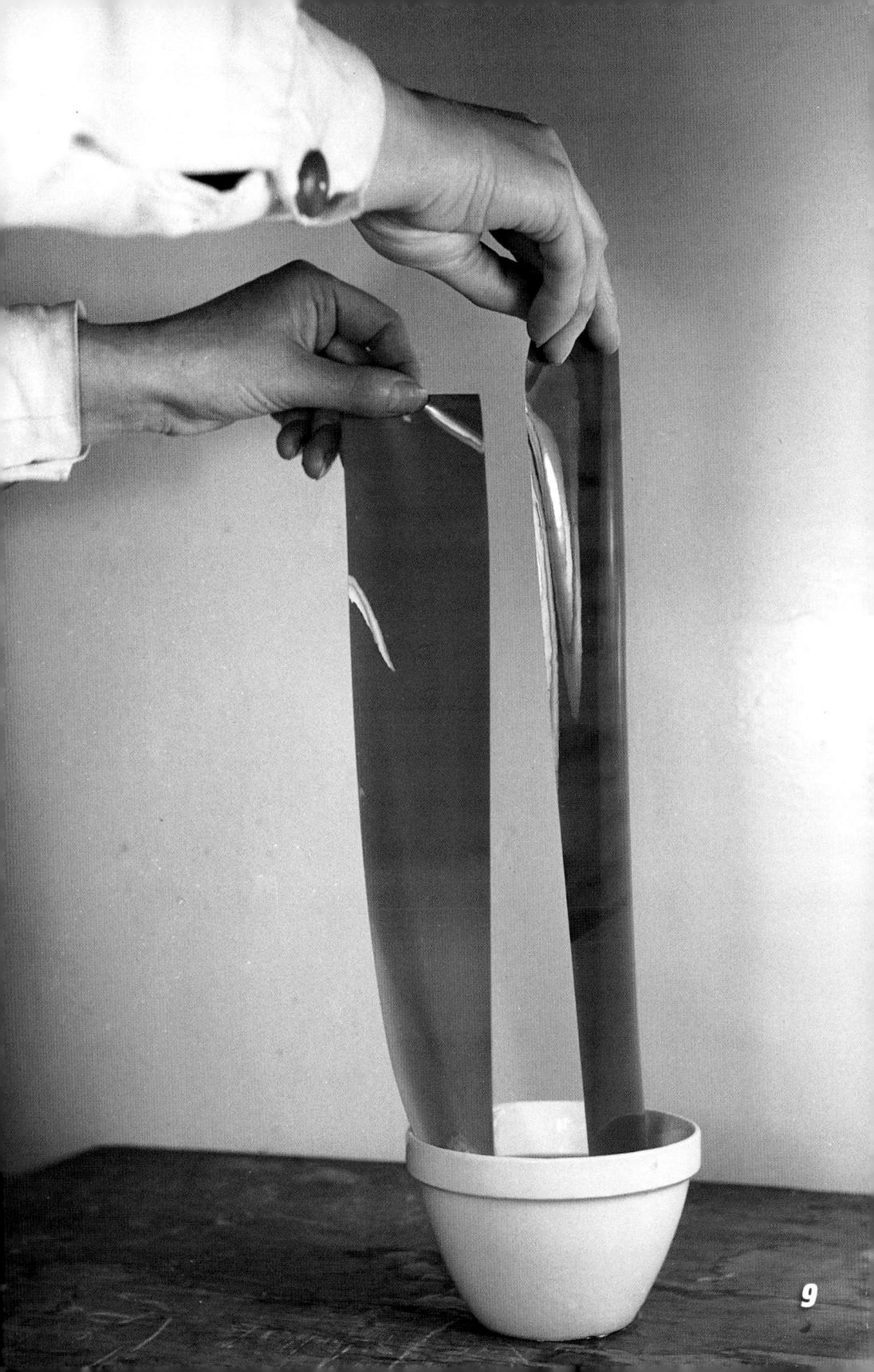

THE DISCOVERY

In 1907, Leo Baekeland introduced Bakelite. It was the first fully-**synthetic** plastic. It was strong, heat resistant, and electrically **insulating**. This discovery opened the doors to new possibilities.

In 1926, inventor Waldo Semon found a way to make plastic flexible. Known as polyvinyl chloride (PVC), it quickly became one of the most widely used products in the world!

More advances in plastic came during **World War II**. It was used as **shock absorbers**, **insulation** for radar cables, and parachutes! After the war, plastics became important in everyday life.

Soon, plastic was everywhere. It was used in packaging, household items, and electronics. Plastic also helped improve industries like healthcare, construction, and transportation.

The 21st century has seen a growing awareness of plastic waste. This has led to increased efforts toward recycling and exploring **sustainable** options.

Plastic has changed business and simplified daily life. While the environmental challenges are still being worked on today, history has certainly been shaped by the discovery!

GLOSSARY

celluloid – a clear flammable plastic that was eventually used to film movies.

insulation – material used to keep something from losing or transferring electricity, heat, or sound.

ivory – a hard, white material usually made from the teeth of animals or elephant tusks.

shock absorber – a device that limits the vibration a vehicle produces.

sustainability – the practice of protecting the natural environment while driving innovation and not compromising the way of life.

synthetic – something that is human-made by a chemical process. Synthetic products include plastic and many kinds of fabrics, dyes, and drugs.

World War II – (1939–1945) a war fought in Europe, Asia, and Africa. Great Britain, France, the United States, the Soviet Union, and their allies were on one side. Germany, Italy, Japan, and their allies were on the other side.

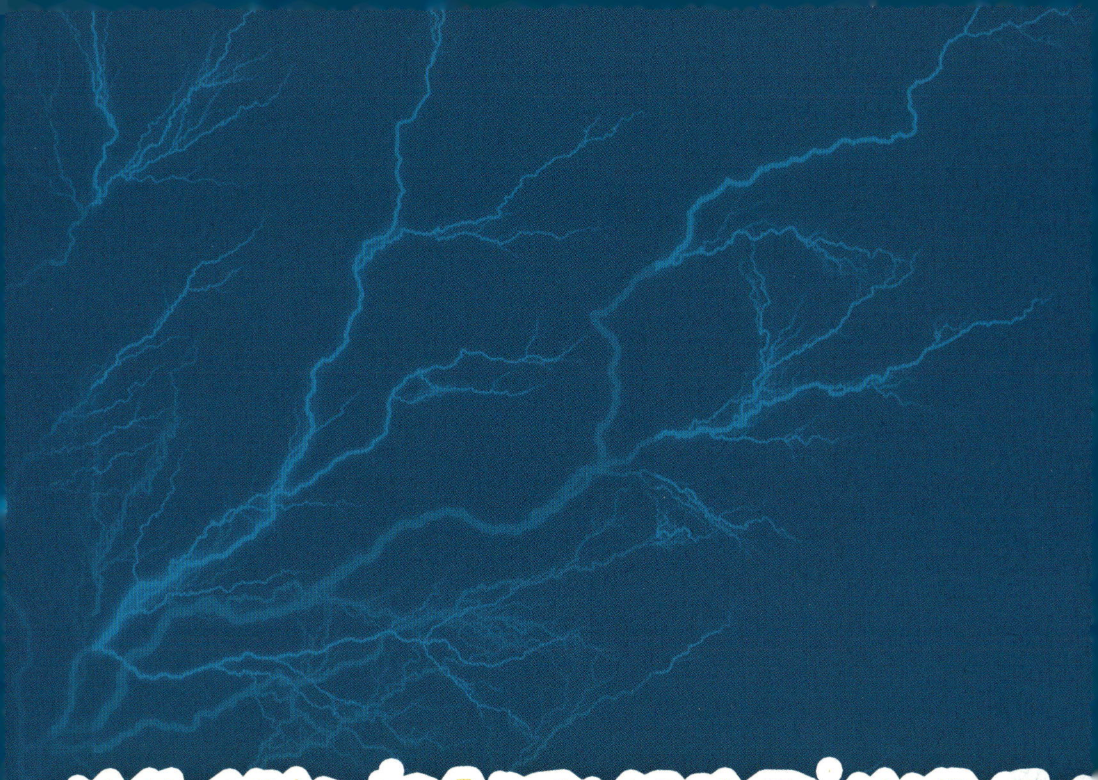

ONLINE RESOURCES

To learn more about plastic, please visit **abdobooklinks.com** or scan this QR code. These links are routinely monitored and updated to provide the most current information available.

INDEX

Baekeland, Leo 11

Bakelite 11

environment 19, 21

Hyatt, John Wesley 6, 8

Polyvinyl chloride (PVC) 13

Semon, Waldo 13

uses 8, 11, 13, 14, 16, 21

waste 19, 21

World War II 14